EINSTEIN'S RELATIVITY

Private Universes

A step by step explanation of how speed affects space and time.

Edward Hickton / Stewart Keach

Text copyright © 2015 Edward Hickton & Stewart Keach

All rights reserved

Published by Resaltos

ISBN: 1517000335
ISBN-13: 978-1517000332

INTRODUCTION

What's special about our book?

Good question!

We use everyday language, and just one example situation, to explain all the space and time aspects of Einstein's Special Theory of Relativity!

There's a "red thread" running through the book – every step is built on the steps taken before. And by starting with something fairly familiar, like "speed equals distance divided by time", and using simple maths, we're able to end up showing why it's impossible to travel faster than the speed of light!

What do we mean by "simple maths"? We mean the maths we're typically taught at school by the age of sixteen! Really – relativity isn't mathematically complex!

So, let's get on with the book, in which we discuss the space and time aspects of Einstein's Special Theory of Relativity, including how we intuitively expect space and time to behave, how they actually behave, why there are differences, what those differences are, what are their consequences, and how to quantify them.

Edward Hickton & Stewart Keach. August, 2015.

'I often say that when you can measure what you are speaking about, and express it in numbers, you know something about it; but when you cannot express it in numbers, your knowledge is of a meagre and unsatisfactory kind; it may be the beginning of knowledge, but you have scarcely, in your thoughts, advanced to the stage of science, whatever the matter may be...'

~ Lord Kelvin, physicist.

'Meten is weten.' (To measure is to know.)

~ Dutch proverb.

'... in mathematics, you don't understand things, you just get used to them'

~ John von Neumann, mathematician.

'Anyone who has never made a mistake has never tried anything new.' 'The important thing is not to stop questioning. Curiosity has its own reason for existing.'

~ Albert Einstein.

CONTENTS

1	WHAT WOULD YOU EXPECT?	1
2	WORKING OUT WHAT DOESN'T HAPPEN	17
3	TRANSFORMING SPACE	33
4	DISCOVERING GAMMA	39
5	TRANSFORMING TIME	44
6	EINSTEIN MEETS PYTHAGORAS	52
7	TIME DILATION	59
8	LENGTH CONTRACTION	71
9	PUTTING IT ALL TOGETHER	79
10	MARS IS CLOSER THAN YOU THINK	96
11	THE UNIVERSAL SPEED LIMIT	100

CHAPETR 1

WHAT WOULD YOU EXPECT?

Stu – Ed, I've heard that moving objects shrink, and moving clocks run slow. I remember hearing that this has got something to do with Einstein, his "Theory of Relativity", the speed of light, "$E=MC^2$", all that kind of stuff. I'd really love to understand more about it all, but I guess that would require a brain like Einstein's, wouldn't it?

Ed – Not at all, Stu! The best way to understand it is to take it step by step, and you don't have to be an Einstein to do that, you just need to follow some logic, and do a little maths. Would you like to give it a go?

Stu – Sure. Where do we start?

Ed – Well, although we want to explain why moving objects shrink and moving clocks run slow, which means we're interested in space and time, let's start instead with energy, and a little detour into that famous equation you mentioned, "$E=MC^2$". What does it tell us, do you think?

Stu – Well, if I remember right, then "E" stands for energy, "M" stands for mass and "C" stands for the speed of light, which is something like 300,000 kilometres per second. So the equation "$E=MC^2$" tells us that "energy equals mass multiplied by the speed of light squared", and when something is "squared", it means it's multiplied by itself.

Ed – That's right. Is there anything else it tells us that you can think of?

Stu – Well, it's an equation, so the two sides must always be in balance. I remember learning at school that the two sides of an equation are like the two sides of a set of weighing scales. The scales only stay in balance if, whenever something changes on one side, an equivalent change is made on the other side. That means that, according to "$E=MC^2$", because the speed of light is fixed, I mean we know it's 300,000 kilometres per second, then the more mass you have, the more energy you'll have, and vice-versa.

Ed – That's right. And do you also remember learning that if you know the values of all the items in an equation except one, then you can use algebra to work out the missing item?

Stu – You mean like working out what "X" equals? Yes, I remember doing that a lot!

Ed – Good. So in "$E=MC^2$", if you know mass, then as you said, since the speed of light has a constant value, you can use algebra to work out energy, and vice-versa. Does that make sense?

Stu – Yes, It would be just like working out what "X" equals, except we'd be working out what "E", or "M", equals.

Ed – Exactly. And since there are three variables, "E" for energy, "M" for mass and "C" for the speed of light, we can write the equation in three different ways - one for energy, one for mass, and one for the speed of light. That's because, like we said just now, we can change one side of the equation so long as we make an equivalent change on the other side. So for "E=MC²" we can divide both sides of the equation by "C²" like this:

$$\frac{E}{C^2} = \frac{MC^2}{C^2}$$

Figure 1a

… and we can split the right-hand side into two parts like this:

$$\frac{E}{C^2} = M \times \frac{C^2}{C^2}$$

Figure 1b

… and "C²" divided by itself equals 1, and "M" times 1 is the same as just "M", so:

$$\frac{E}{C^2} = M$$

Figure 1c

… and swapping the two sides around we get:

$$M = \frac{E}{C^2}$$

Figure 1d

Stu – I follow that. Am I right in remembering that there's a shortcut for those steps?

Ed – That's right. It's called "cross-multiplying". When you have an equation that includes fractions, you can move a variable on one side of the equation to the other side, by changing it from being on the top half of the fraction (what's known as a "numerator"), to the bottom half (to become a "denominator"), and vice-versa.

Stu – But "E=MC^2" doesn't have any fractions in it!

Ed – Well yes, and no. Don't forget that "E=MC^2" can also be written as:

$$\frac{E}{1} = \frac{MC^2}{1}$$

Figure 2a

Stu – I see. So just now, "C^2" was swapped from being a numerator on the right-hand side of the equation to become a denominator on the left-hand side like this:

$$\frac{E}{1C^2} = \frac{M}{1}$$

Figure 2b

… and since "1C^2" is the same thing as just "C^2", and "M over 1" is the same thing as just "M" like you said before, the result is:

$$\frac{E}{C^2} = M$$

Figure 2c

… which you then swapped around to put "M" first.

Ed – That's right. And we can do the same thing with the "M" in "E=MC²" to get:

$$C^2 = \frac{E}{M}$$

Figure 2d

… and when we look at this way of writing the equation, we can see "C squared" - which, since "C" itself is fixed, must also be fixed - as a kind of ratio, or exchange rate, between energy and mass!

Stu – You mean like in currency exchange? But instead of say, euros and dollars, we've got energy and mass?

Ed – Yes, that's it! So energy equals mass multiplied by the exchange rate, and mass equals energy divided by the exchange rate.

Stu – That's interesting - I've never thought of "E=MC²" like that before. It seems that the speed of light must be very important, if it's some kind of a link between things as fundamental as energy and mass!

Ed – Indeed! We may tend to think of the speed of things as nothing more than a ratio, like in "kilometres per second", or "miles per hour" - a ratio that can have any value, and that has no limit. But actually it's much more than that, because speed *does* have a limit, and that limit is a fundamental property of the universe. We'll come back to that in a moment, but for now we'll end our little detour into energy, and return to our main interests, i.e., space and time. You may remember that, just as "$E=MC^2$" tells us that energy and mass are related to speed - the speed of light, as we've just been talking about - time is also related to speed?

Stu – Yes, if I've got it right, then speed equals distance divided by time, which is why we say for example, "kilometres per second". It's a ratio, like you said just now - the ratio between a certain distance travelled, and the amount of time it takes.

Ed – Good. Now, have you heard that nothing travels faster than the speed of light?

Stu – Yes, I have. Why is that, by the way?

Ed – Well, light weighs nothing, and so it travels at the fastest possible speed.

Stu – Makes sense! So that must mean that the speed of light is equal to the speed limit of the universe that you just mentioned just now?

Ed – Absolutely right! Now, did you ever wonder whether the speed of light is absolute, or relative?

Stu – No, and I'm not sure what you mean by that!

Ed – It's not an obvious question, but it's a very important one, and it's related to the word "relativity" in Einstein's famous theory. If the speed of light is absolute, then it never changes. But if the speed of light is relative, then it varies according to what it's moving in relation, or relative, to.

Stu – I think I get it. Can you give me an example?

Ed – Sure. As you already said, the speed of light is around 300,000 kilometres per second. Now, imagine the following situation. You're sitting on a train, and the train is standing at a station. I'm standing on the station platform. You're holding a torch, and pointing it in the direction the train will travel when it leaves the station. While the train is standing at the station, the light will shine out of your torch at 300,000 kilometres per second, won't it?

Stu – Yes, because that's the speed of light!

Ed – Indeed. Now, imagine that the train pulls out of the station at 0.1 kilometres per second. Do you think that I, standing on the platform, would see the light shine out of your torch (a) faster, at 300,000 + 0.1 = 300,000.1 kilometres per second, or (b) unchanged, at 300,000 kilometres per second?

Stu – I'd expect (a), as the light would have been sped up by the motion of the train. Just like if I walked down one of the carriages in the same direction as the train is travelling - from the platform you'd see me going passed at my walking speed *plus* the train speed.

Ed – Yes, intuitively you'd expect so, which would mean that the speed of light is relative, not absolute. Relative to *you*, travelling at 0.1 kilometres per second on the train holding the torch, the speed of light would be unchanged at 300,000 kilometres per second. But relative to *me*, standing on the platform watching the train go by, it would be travelling faster at 300,000.1 kilometres per second.

Stu – Yes, that's what I'd expect.

Ed – Actually, the speed of light is absolute, and we'd both see the light travel at 300,000 kilometres per second, not only when the train is standing at the station, but also when it's moving!

Stu – How can that be?

Ed – Well, when you think about it, it makes sense, because nothing can travel faster than the speed of light. Imagine that the train is travelling incredibly fast at say, 90% of the speed of light, that's 0.9 times 300,000 equals 270,000 kilometres per second. And imagine that you run from one end of the carriage to the other, also at 90% of the speed of light. Then you'd intuitively expect that I, standing on the platform, would see you going passed at 270,000 + 270,000 = 540,000 kilometres per second, which would be faster than the speed of light, which is impossible, because nothing can travel faster than the speed of light. Now, we both know it's impossible for either you or the train to move that fast, but you see the point, don't you?

Stu – The point being that, for the speed of light to really be a *maximum* speed, it must be an *absolute* speed, and not a *relative* speed?

Ed – Spot on! It can't be an absolute maximum if it varies. As we said earlier, speed isn't just a ratio that can have any value, it has an absolute limit that's a fundamental property of the universe!

Stu – OK! But if I were to run at 270,000 kilometres per second in a train carriage travelling at 270,000 kilometres per second, and you say it's impossible for you to see me going passed you at 540,000 kilometres per second, then at what speed would you see me going passed you?

Ed – That's a good question! But we're not ready to answer it yet. Don't worry - we'll be able to answer it later!

Stu – OK. But how can we be sure that the speed of light is absolute, I mean, has it been proved?

Ed – Yes, many observations have proved it. A famous example is the Michelson-Morley experiments of 1887, which showed that the speed of light is the same when it travels, not only in the same direction as the Earth's rotation, but also at right angles to it.

Stu – Meaning that light isn't sped up by the rotation of the earth?

Ed – Exactly. So, whether a source of light is moving towards you or away from you, or not moving at all, the light still travels at 300,000 kilometres per second, completely contrary to what you'd expect!

Stu – So, going back to our example of me with the torch on the train, and our intuition telling us that the speed of light would be increased for you on the platform due to the motion of the train, we're now saying that doesn't happen?

Ed – That's right. Like I said, we'd both experience the same speed for the light shining from your torch, both when the train is standing at the station, and when it's moving!

Stu – But earlier we said that speed is a ratio calculated by the equation "speed equals distance over time". And we also said that, for any equation, both sides must always be equal. Are we now saying that, for the speed of light, the value for the speed side of the equation, which we intuitively thought would increase from your point of view due to the motion of the train, doesn't increase after all?

Ed – That's right. Intuitively you'd think that the motion of the train causes the speed of light to increase, but that's not what happens.

Stu – So what does happen then?

Ed – Well, as Sherlock Holmes famously said, "When you have eliminated the impossible, whatever remains, however improbable, must be the truth." So, if the speed side of the "speed equals distance over time" equation can't change to compensate for the motion of the train so that we each experience a different speed for light, then in order for the two sides of the equation to stay in balance, the other side of the equation must change, i.e., the side with distance and time!

Stu – You mean that, in order for us to experience the *same* speed for light, we must experience a *different* time period for a second, or a *different* distance for a kilometre?

Ed – Precisely! We intuitively expect the speed of light to be relative and experienced differently by each of us, and distance and time to be absolute, and experienced identically. In fact, it's totally the other way around! The speed of light is absolute and experienced identically, while distance and time are relative, and experienced differently!

Stu – Wow! In our example of the torch and the train, what difference will the train's speed make to the way we experience distance and time?

Ed – Well, for me, as I watch the train pull out of the station, its length will shrink, and time on the train will slow down. And the faster the train goes, the more these things will happen!

Stu – That sounds like the same kind of thing as what we're most interested in, i.e., distance shrinking and clocks running slow.

Ed – Indeed! And moreover, it all happens in such a way that, no matter at what speed the train is travelling, even though we experience distance, i.e., space, and time differently, dividing distance by time will always give the same value for the speed of light for both of us. We'll do the maths for that later - we're not quite ready for it yet.

Ed – OK. So is there some kind of formula for working out by how much time slows down and distance shrinks?

Ed – Indeed there is, and it's called the "Lorentz Factor", after the Dutch physicist Hendrik Lorentz who derived the equations used by Einstein in his famous "Theory of Relativity". Let's explore it all further by introducing the idea of what's known in scientific circles as a "frame of reference", or as you and I might prefer to think of it, our own "private universe".

Stu – I like the idea of my own private universe! What's in it?

Ed – Everything that's not moving, by which I mean, not moving relative to you. So, in the example of the train and the torch, the train and the torch are in your private universe because they aren't moving relative to you, since you and the train and the torch are all moving together. But I, standing on the platform, am not in your private universe, because I'm moving relative to you, because we're moving apart.

Stu – OK, I get that. So then other things like the platform, and the railway line that the train is travelling on, aren't in my private universe either?

Ed – That's right. Now, let's imagine a situation which demonstrates how you and I could experience time differently!

Stu – Sounds interesting!

Ed – Indeed! Apparently Einstein used to do this kind of thing quiet a lot, and called it a "thought experiment". So your brain is about to follow a thought process just like the kind of thinking the great man himself used to do.

Stu – I hope my brain is up to it!

Ed – Don't worry, just because Einstein thought it, it doesn't mean it has to be rocket science! Now, our imaginary situation demonstrates how you and I could experience time differently in a way that's totally independent of what time each of us thinks it is, according to our watch, phone, laptop, or whatever it is that we're using to tell the time.

Stu – But how is it possible for us to know that we're experiencing time differently without considering what time it is?

Ed – Good question! We'll do it by considering two events. And we'll see that these two events happen simultaneously for you, but at different times for me!

Stu – That sounds weird!

Ed – Sure does! Now, let's go back to our example of the train and the torch again. Let's imagine your train is on its way somewhere, and it doesn't stop at my station – so it's going to go passed me standing on the platform. Let's also imagine that you're sitting exactly halfway down your railway carriage, so that the distance between you and both the front and the back of the carriage is exactly the same. Also, let's swap your torch for a lantern from which light shines out in all directions. Finally, let's imagine that, at each end of the carriage, there's a door controlled by a photo-electric cell, so that when the light from your lantern shines on the cell, the door opens.

Stu – OK, I think I can imagine all that!

Ed – Good! Let's say that, to start with, your lantern is switched off, and that you switch it on at the exact moment that you pass me standing on the platform. Now, what would happen when you switch your lantern on?

Stu – Well, I guess the light would shine out from the lantern in all directions, and reach both ends of the carriage, where it would hit each of the two photo-electric cells, triggering the opening of both doors.

Ed – That's right. And because you're sitting exactly halfway down the carriage, the light would have to travel exactly the same distance to each door, so it would take exactly the same time to reach each door, so the doors would open at exactly the same time.

Stu – Yes, the doors would open simultaneously.

Ed – Yes, in your private universe that would be the case. But now let's consider how things will be for me in my private universe, standing on the platform, watching the train go by.

Stu – Oh yes, I forgot that we're in different private universes. What difference does it make?

Ed – Well, as strange as it may seem, it means that I wouldn't see the doors opening simultaneously. I'd see them opening at different times!

Stu – But how could that be?

Ed – Well, in between you switching on the lantern, and the light reaching the doors, the train has moved, hasn't it?

Stu – Yes, not by much, but a little.

Ed – So the back door of the train has moved towards the light, and the front door has moved away from the light.

Stu – Yes, that follows.

Ed – So the light has less distance to travel to the back door than the front door, and since it always travels at the same speed of 300,000 kilometres per second, then it will reach the back door before the front door, so I'll see the back door open before the front door!

Stu – So I see the doors open at the same time, but you see them open at different times!

Ed – Yes! Because just like we talked about earlier, the speed of light isn't changed by the speed of the train. You'd intuitively expect the light travelling to the front door to be sped up by the moving train, which would compensate for the extra distance it has to travel, and vice-versa for the back door. Then the doors would open simultaneously for me, just like they do for you. But that's not what happens!

Stu – I see! So what's happening to time? I mean, does it have something to do with the slowing down of time that we talked about earlier?

Ed – The key thing here isn't a slowing down of time – although that will be happening too - it's a "shifting" of time – by which I mean that the moment at which I experience an event is shifted backwards or forwards compared to the moment at which you experience it.

Stu – You mean that, because the speed of light isn't "shifted", i.e., sped-up, by the speed of the train, then the times at which the doors open is shifted instead?

Ed – You've summed it up very well there, Stu!

CHAPETR 2

WORKING OUT WHAT DOESN'T HAPPEN

Ed – We can do some maths to demonstrate, step by step, what *intuitively* happens, and later we can update our maths, again step by step, so that it tells us what *actually* happens.

Stu – Simple maths I hope!

Ed – Don't worry, just the sort of algebra we did at school, all based on the formula we mentioned earlier, of "speed equals distance over time". And after we've updated our maths so that it deals with what *actually* happens, then it will cover everything, i.e., time shifting, time dilation, length contraction - even why it's impossible to go faster than the speed of light!

Stu – Sounds good!

Ed – Indeed! And for *any* event that takes place at a given time and place in one of our private universes, our maths will enable us to work out at what time and place that same event happens in the other's private universe – those are what's known as "Lorentz transformations".

Stu – Seems like thought experiments can be pretty useful things! Let's get started with our maths!

Ed – OK. Let's start with the formula for speed:

$$\textbf{Speed} = \frac{\textbf{Distance}}{\textbf{Time}}$$

Figure 3a

Stu – I remember that one!

Ed – Good. Now, when we know the value of any two of speed, distance and time, we can use algebra to work out the third one, because, just like we did with "E=MC²" by cross-multiplying, the formula can be arranged in any of three different ways like this:

$$\textbf{Speed} = \frac{\textbf{Distance}}{\textbf{Time}}$$

$$\textbf{Distance} = \textbf{Speed} \times \textbf{Time}$$

$$\textbf{Time} = \frac{\textbf{Distance}}{\textbf{Speed}}$$

Figure 3b

… so because "speed equals distance over time", then "distance equals speed times time", and "time equals distance over speed".

Stu – And that's all we need?

Ed – Yes! Now let's start doing the maths for what intuitively happens.

Stu – Sounds good, let's go.

Ed – OK. As we said, we'll take it step by step. So here's step one. As we just talked about, intuitively we'd expect me standing on the platform to experience the speed at which the light travels to the doors to have been changed due to the motion of the train. What would you expect that change to be?

Stu – Well, I'd expect you to experience the light travelling to the front door to be the speed of light, "C", plus the speed of the train. And for the back door I'd expect the opposite, that is, "C", minus the speed of the train.

Ed – Yes, that's what you'd intuitively expect. Let's follow mathematical tradition and use the letter "V" to represent the velocity, or speed, of the train. Also, let's call you on the train the "Passenger", and me standing on the platform the "Bystander", and we'll keep track of our steps in a table like this, with one column for your private universe, and another column for mine:

		Stu Passenger	Ed Bystander
Light Speed	The speed of light to the front door	(1) C	(1) $(C + V)$
	The speed of light to the back door	(1) $-C$	(1) $(-C + V)$

Figure 4

Stu – OK. But why did you put a "minus" sign in front of the speed of light to the back door for both of us?

Ed – It's because the light is travelling in the opposite direction. So if the speed is positive for the front door, then it's negative for the back door.

Stu – OK, I get it. For me on the train the speed of light is unchanged, so it's always plus, or minus, "C." But for you we'd intuitively expect it to have been changed by the speed of the train, so it will be plus, or minus, "C", plus "V".

Ed – That's right. Now, for step two, let's draw a diagram to illustrate the situation. Also, let's call the distance between you and each door "X" with a little "*m*", or "X^m", and again, like we did just now for speed, we'll make it positive in the forward direction, and negative going backwards:

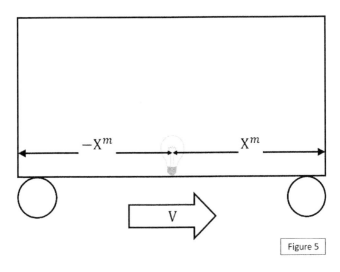

Figure 5

Stu – Why are we adding a little "*m*" to the distances, and not just calling them plus, or minus, "X"?

Ed – It's because, as we know, according the theory of relativity, people in different private universes experience space and time differently. So to distinguish the way you experience things from the way I do, we'll add a little "*m*" for distances and times in your private universe. The "*m*" is for "motion", because your private universe is moving relative to mine.

Stu – OK, I follow that. And I see there's also an arrow in our diagram with a "V" in it, to show the velocity and direction of the train.

Stu – That's right. Now, for you the light has to travel plus, or minus, "X^m" to each door to open it. It's the same distance to each door because, as we said, you're sitting exactly half way down the carriage. Bur for me in my private universe it will be different because, like we said earlier, the train will have moved a bit in the time between the light leaving the lantern and reaching the doors.

Stu – Yes, like we said, for you the front door will have moved away from the lantern during the time it takes the light to reach it, and the back door will have moved towards the lantern.

Ed – That's right, but we haven't worked out by how much the train will have moved yet, so we'll put question marks in our table for how far the light has to travel to each door in my private universe for now, until we get to the step where we work it out:

		Stu Passenger	Ed Bystander
Light Speed	The speed of light to the front door	**C**	**(C + V)**
	The speed of light to the back door	**−C**	**(−C + V)**
Light Distance	The distance the light has to travel to the front door	②X^m	②?
	The distance the light has to travel to the back door	②$-X^m$	②?

Figure 6

Stu – OK. When will we be able to work out what the

actual values for those question marks are?

Ed – Quite soon. But we can't do it just yet because, as we know, in order to work out distance we need to know both speed and time. We already know that the speed of the train is "V", but we haven't yet worked out the time between the lantern being switched on and the doors opening in your private universe on the train. So let's make that step three – let's call that time "T^m". What do you think it will be for the front door?

Stu – Well, we know that "time equals distance over speed". The speed of light is "C", and the distance the light has to travel to the front door is "X^m". So the time between the lantern being switched on and me seeing the front door open will be:

$$T^m = \frac{\text{Distance}}{\text{Speed}}$$

$$T^m = \frac{X^m}{C}$$

Figure 7a

Ed – That's right. And what about the back door?

Stu – Well, that would be the distance the light has to travel to the back door, i.e., "minus X^m", divided by the speed of light, "minus C", so the time would be:

$$T^m = \frac{\text{Distance}}{\text{Speed}}$$

$$T^m = \frac{-X^m}{-C}$$

Figure 7b

Ed – Good. And do you remember from school that a "minus" divided by a "minus" is a "plus"?

Stu – Yes, I do. So the answer is the same as for the front door, as we'd expect, i.e:

$$T^m = \frac{X^m}{C} \qquad \text{(3)}$$

From Figure 7a

Ed – Indeed. Let's update our table:

		Stu Passenger	Ed Bystander
Light Speed	The speed of light to the front door	C	$(C + V)$
	The speed of light to the back door	$-C$	$(-C + V)$
Light Distance	The distance the light has to travel to the front door	X^m	?
	The distance the light has to travel to the back door	$-X^m$	?
Time	The time elapsed before the front door opens	(3) $T^m = \frac{X^m}{C}$	(3) ?
	The time elapsed before the back door opens		(3) ?

Figure 8

Stu – OK. And I see you've put question marks again for that in your column, until we've worked it out.

Ed – That's right. So now we can work out the time it takes for the doors to open for you in your private universe. If, for example, your carriage is 6 metres long, then each door would be 3 metres away from you, and knowing that a metre is one thousandth of a kilometre, then it would take:

$$T^m = \frac{X^m}{C} \quad (3)$$

$$T^m = X^m \div C$$

$$T^m = \frac{3}{1,000} \div 300,000$$

$$T^m = \frac{3}{1,000} \times \frac{1}{300,000}$$

$$T^m = \frac{3}{300,000,000}$$

$$T^m = \frac{1}{100,000,000}$$

$$T^m = 10 \text{ nanoseconds}$$

Figure 9

Stu – How long is a nanosecond?

Ed – A nanosecond is a thousand millionth of a second.

Stu – That's not long! Light travels pretty fast doesn't it?

Ed – Indeed it does!

Stu – OK, I get that. Now can we start working out the time we'd intuitively expect it takes for the doors to open in your private universe?

Ed – Let's try. Same as always, to work out time we need to know both distance and speed. Speed we already worked out in step one as "(C + V)" for the front door, and "(– C + V)" for the back door. Now we have to work out the distance. As we said earlier, the train will have travelled some distance between you switching on the lantern and the light reaching the doors, and we need to work out what that distance is. Let's illustrate that by including a question mark for it in our diagram:

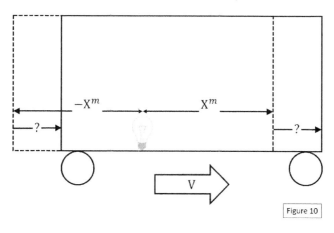

Figure 10

Stu – OK. How can we work that distance out?

Ed – We can use our speed, distance and time formulas again. Do you think you can do it?

Stu – Let me try. To work out distance we need to know speed and time. We know the speed of the train is "V", and we called the time it takes for the doors to open for me on the train to be "T^m". So the distance travelled by the train during the time it takes for the light to reach the doors will be:

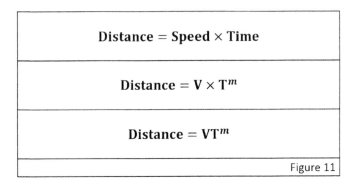

Figure 11

Ed – Good. We need that information so that we can complete step four, which is to work out how far, for me in my private universe on the platform, the light has to travel for the front door to open. Since I'm not moving we'll call that "X", without a little "m". Can you do that?

Stu – Well, for the light to reach the door it has to travel half the carriage length, i.e., "X^m", plus the distance, "VT^m", that we just worked out that the train travels between me switching on the lantern and the light reaching the front door, so:

$$X = X^m + VT^m \qquad (4)$$

Figure 12

Ed – Good. Now let's work out the distance the light has to travel in the opposite direction for the back door to open for me – can you do that too?

Stu – That would again be half the carriage length, but this time with a "minus" sign since it's in the opposite direction to the front door, adjusted by the distance that the train travels between the lantern being switched on and the light reaching the door, so:

$$-X = -X^m + VT^m \qquad \text{\textcircled{4}}$$

Figure 13

Ed – Good. Let's add those two new results to our table:

		Stu Passenger	Ed Bystander
Light Speed	The speed of light to the front door	C	$(C + V)$
	The speed of light to the back door	$-C$	$(-C + V)$
Light Distance	The distance the light has to travel to the front door	X^m	$X^m + VT^m$ \quad \textcircled{4}
	The distance the light has to travel to the back door	$-X^m$	$-X^m + VT^m$
Time	The time elapsed before the front door opens	$T^m = \dfrac{X^m}{C}$	?
	The time elapsed before the back door opens		?

Figure 14

Stu – Looks like we're almost there - only two question marks left!

Ed – Indeed! Let's finish up with step five. How long do you think it will take for the front door to open for me, i.e., what will "T" be?

Stu – Well, to work out time we need to know speed and distance. We worked out the speed to be "$(C + V)$" earlier, and we just calculated the distance, so:

$$T = \frac{\text{Distance}}{\text{Speed}}$$

$$T = (X^m + VT^m) \div (C + V)$$

Figure 15a

Ed – Yes. And remember that earlier we worked out that:

$$T^m = \frac{X^m}{C} \qquad (3)$$

From Figure 7a

Stu – OK. So if we substitute "X^m over C" for "T^m", then:

$$T = \left[X^m + \left(V \times \frac{X^m}{C}\right)\right] \div (C + V)$$

$$T = \left(X^m + \frac{VX^m}{C}\right) \div (C + V)$$

$$T = X^m\left(1 + \frac{V}{C}\right) \div (C + V)$$

$$T = X^m\left(1 + \frac{V}{C}\right) \times \left(\frac{1}{C + V}\right)$$

Figure 15b

… and multiplying that out, i.e., each of the terms in the first bracket by the term in the second, then:

$$T = X^m \left[\left(1 \times \frac{1}{C+V}\right) + \left(\frac{V}{C} \times \frac{1}{C+V}\right)\right]$$

$$T = X^m \left(\frac{1}{C+V} + \frac{V}{C(C+V)}\right)$$

Figure 15c

… and multiplying both the top and bottom of the first term in the bracket, i.e., "one over C + V" by "C", so that it has the same denominator as the second one, so that they can be added together, then:

$$T = X^m \left(\frac{1C}{C(C+V)} + \frac{V}{C(C+V)}\right)$$

$$T = X^m \left(\frac{C}{C(C+V)} + \frac{V}{C(C+V)}\right)$$

$$T = X^m \left(\frac{C+V}{C(C+V)}\right)$$

$$T = X^m \left(\frac{1}{C}\right)$$

$$T = \frac{X^m}{C} \qquad (5)$$

Figure 15d

Ed – Well done Stu! Now, do you notice anything about that result?

Stu – Yes. It's exactly the same as the result we worked out in step three for me in my private universe!

Ed – That's right! It's what you'd intuitively expect, which is that for me the increase in the speed of light, and the increased distance the light has to travel, cancel each other out. So the front door opens for me at the same time as for you. Let's finish up by working it out for the back door too.

Stu – That would be the same as before, but with "$-X^m + VT^m$" instead of "$X^m + VT^m$" for distance, and "$(-C + V)$" instead of "$(C + V)$" for speed, so:

$$T = \frac{\text{Distance}}{\text{Speed}}$$

$$T = (-X^m + VT^m) \div (-C + V)$$

$$T = \left(-X^m + \frac{VX^m}{C}\right) \div (-C + V)$$

$$T = -X^m\left(1 - \frac{V}{C}\right) \times \left(\frac{1}{(-C + V)}\right)$$

$$T = -X^m\left[\left(1 \times \frac{1}{(-C + V)}\right) - \left(\frac{V}{C} \times \frac{1}{(-C + V)}\right)\right]$$

$$T = -X^m\left(\frac{1}{(-C + V)} - \frac{V}{C(-C + V)}\right)$$

$$T = -X^m\left(\frac{1C}{C(-C + V)} - \frac{V}{C(-C + V)}\right)$$

$$T = -X^m\left(\frac{C - V}{C(-C + V)}\right)$$

$$T = -X^m\left(\frac{C - V}{-CC + CV}\right)$$

$$T = -X^m \left(\frac{C - V}{-C(C - V)} \right)$$

$$T = -X^m \left(\frac{1}{-C} \right)$$

$$T = \frac{-X^m}{-C}$$

$$T = \frac{X^m}{C} \quad \text{(5)}$$

Figure 16

Ed – That's right! And that's exactly the same result that we calculated for the front door for me, and also for both doors for you! Let's complete our table:

		Stu Passenger	Ed Bystander
Light Speed	The speed of light to the front door	C	$(C + V)$
	The speed of light to the back door	$-C$	$(-C + V)$
Light Distance	The distance the light has to travel to the front door	X^m	$X^m + VT^m$
	The distance the light has to travel to the back door	$-X^m$	$-X^m + VT^m$
Time	The time elapsed before the front door opens	③	$T^m = T = \dfrac{X^m}{C}$
	The time elapsed before the back door opens	⑤	

Figure 17

Stu – So the maths is confirming what we'd intuitively expect – the doors open 10 nanoseconds after I turn on the lantern for both of us?

Ed – Yes, and it's all because we started with the premise that the speed of light is relative, not absolute. But as we said earlier, that's not correct. If we want to work out what *actually* happens, we have to work on the premise, or rather the fact, that the speed of light is absolute.

CHAPTER 3 – TRANSFORMING SPACE

Stu – You mean like we said earlier about the speed of the light from my torch not being increased by the speed of the train?

Ed – Exactly. Now let's ask ourselves this – how can it be that I as bystander don't experience any increase in the speed of the light on the moving train?

Stu – Well, could it be that for you the light is taking more time to travel a given distance than for me? I mean, your stationary watch is running faster than my moving one?

Ed – That would make sense.

Stu – Yes, it reminds me of us experiencing the doors opening at different times - it seems something strange is going on with time.

Ed – Indeed! But couldn't it also be that the light is travelling a lesser distance in a given time, for example, what you measure as one metre I measure as a lesser distance?

Stu – Yes, that's also a logical explanation. And if we suspect that something strange is going on with time, I guess we can't rule out that something strange might be going on with distance too!

Ed – Indeed! And, in fact, we can already conclude that

something strange *is* going on with distance!

Stu – Really? Why's that?

Ed – Well, it's because if you want to measure the length of a moving object, you have to measure both ends at the same time.

Stu – You mean that, if you measure each end of something that's moving at different times, because the object will have moved in between the two measurements, you'll get the wrong length?

Ed – Precisely! The measurement of each end must be simultaneous events. And we already know from our thought experiment that events that are simultaneous for you are not simultaneous for me!

Stu – So we'd measure things differently?

Ed – Exactly!

Stu – Interesting! So how are we going to proceed? I mean, how are we going to figure out what's going on, and to quantify the time and distance differences between our private universes so that we can work out when the doors open for you in your private universe?

Ed – Actually, with what we've done up to now we already have almost everything we need to do that. There's just one piece of information we miss.

Stu – Only one?

Ed – Yes. And we can figure it out by thinking about this for a moment - have you ever been sitting on a train waiting for it to leave the station, and been fooled into thinking that your train has started moving, until you realise that it's the train next to yours that's moving in the opposite direction, and yours is still standing still?

Stu – Yes, that's a weird feeling!

Ed – It is, isn't it? It shows that experiencing moving passed something that's stationary is indistinguishable from experiencing being stationary while something moves passed you. You can only figure out what's really moving by looking at something else, like the platform for example, that you think is stationary.

Stu – Why do you say that you "think" the platform is stationary? Surely it *is* stationary!

Ed – Well, the earth is spinning at several hundred kilometres an hour.

Stu – Well yes, the earth is spinning on its axis. So it's really the centre of the earth that's stationary I suppose.

Ed – Well, no, because the earth is orbiting the sun at over a hundred thousand kilometres an hour. And the sun isn't stationary either, because our galaxy is spinning at about eight hundred thousand kilometres an hour. And our galaxy isn't stationary either, because our universe is expanding…

Stu – So the station platform is moving at one speed relative to the centre of the earth, another speed relative to the sun, a different speed again relative to the centre of the Milky Way, and even a different speed again relative to say, Andromeda?

Ed – That's right! But we'd perceive it as stationary!

Stu – So is anything really stationary? It seems that everything is moving at lots of different speeds all at the same time!

Ed – Exactly! It all depends on what you compare with what.

Stu – So we can only say we're moving, or not moving, by comparing our own private universe with another one? So being stationary, or moving, is relative, not absolute?

Ed – Precisely! If someone asks you "how fast are you going", then it would be perfectly valid to answer "relative to what?"

Stu – So everything in my private universe I'd call stationary. But someone in another private universe could say that everything in their private universe is stationary, and it's my private universe that's moving!

Ed – That's right!

Stu – Fascinating! But how is that relevant to our thought experiment?

Ed – It tells us something very important - it tells us that relativity is *symmetrical*.

Stu – OK, and how does that help us?

Ed – It means that, whatever adjustments I'd have to make to distances and times in my private universe to find out what those distances and times are for you, you'd have to make the *same* adjustments to your distances and times to find out what they are for me!

Stu – You mean that for example, you could say my watch

is running slower than yours, but I could just as well say that your watch is running slower than mine?

Ed – That's right! And it's through knowing that relativity is symmetrical that we're able to work out the formulas to quantify the distance and time differences between our private universes!

Stu – OK - let's do it!

Ed – OK! First of all, let's go back to our thought experiment and what we said we'd intuitively expect to be how far, for me in my private universe on the platform, the light from your lantern has to travel for the front door to open. We said the formula would be:

$$X = X^m + VT^m \qquad (4)$$

Figure 12

Stu – Yes, I remember that. It's half the carriage length, i.e., "X^m", plus the distance, "VT^m", that the train travels between me switching on the lantern and the light reaching the front door.

Ed – That's right. Now, we just talked about adjustments that would have to be made in order to transform distances and times between our private universes. So let's invent a factor for such an adjustment, and follow tradition by calling it "gamma", which has the symbol "γ".

Stu – OK.

Ed – So then we can say that the distance the light has to travel in my private universe is equal to the distance in your private universe, multiplied by our adjustment factor, i.e., gamma. So for step six the formula will become:

$$X = \gamma(X^m + VT^m) \qquad (6)$$

Figure 18

Stu – Makes sense.

Ed – Good. And that formula is known as the "Lorentz transformation" for distance!

Stu – Ah yes, we talked about that earlier. We said that, for any event that takes place at a certain place and time in one of our private universes, the Lorentz transformations will enable us to work out at what time and place that same event happens in the other's private universe.

Ed – That's right! So with the transformation we've just worked out, if we know where something happens in your private universe, and we know what gamma equals, then we'll be able to work out where it happens in mine!

CHAPTER 4 – DISCOVERING GAMMA

Ed – Now, remember that we said that the time between the lantern being switched on, and you seeing each door open, would be the distance the light has to travel to the door, i.e., "X^m", divided by the speed of light, "C", i.e:

$$T^m = \frac{X^m}{C} \qquad (3)$$

From Figure 7a

Stu – Yes, I remember that.

Ed – And just now in step six we said that:

$$X = \gamma(X^m + VT^m) \qquad (6)$$

Figure 18

… so, just like we did earlier, we can substitute "T^m" with "X^m over C" to get:

$$X = \gamma\left(X^m + \frac{VX^m}{C}\right)$$

Figure 19

Stu – OK – I get that.

Ed – Good. Now, with that formula we're looking at things from my point of view. So let's again go back to step four and start creating the reverse of it, i.e., its "mirror image", so that we can also look at things form your point of view.

Stu – OK. In step four we said that for you the distance the light travels to the front door is equal to the distance the light travels to the front door for me, plus the distance the train has travels between the lantern being switched on and the light reaching the door, i.e:

$$\mathbf{X = X^m + VT^m} \qquad ④$$

Figure 12

Ed – That's right. So from *your* point of view the distance the light travels to the front door is equal to the distance the light travels to the front door for *me, minus* the distance the train travels between the lantern being switched on and the light reaching the door, i.e:

$$\mathbf{X^m = X - VT}$$

Figure 20

Stu – Makes sense.

Ed – And again, because relativity is symmetrical, for any distance that I measure in my private universe you, like me, must also multiply it by gamma to find out what you'd experience that distance to be. So, for step seven, here's the Lorentz transformation for distance from your point of view:

$$X^m = \gamma(X - VT) \qquad (7)$$

Figure 21

Stu – OK.

Ed – Now, in the same way as we just said for you, the time between the lantern being switched on and me seeing the front door open would be the distance the light has to travel to the front door, i.e., "X", divided by the speed of light, "C":

$$T = \frac{X}{C}$$

Figure 22

… and, in the same way as before, we can substitute "T" for "X over C":

$$X^m = \gamma\left(X - \frac{VX}{C}\right)$$

Figure 23

Stu – OK, got that.

Ed – So now we have two formulas:

$$X = \gamma\left(X^m + \frac{VX^m}{C}\right)$$	$$X^m = \gamma\left(X - \frac{VX}{C}\right)$$
Figure 19	Figure 23

… and here comes the clever bit. By multiplying them together we can find the formula for gamma!

Stu – Can we do that? I mean, is it mathematically valid to multiply two formulas together?

Ed – Sure! Earlier we talked about equations being like weighing scales that are in balance. What we're about to do now is to multiply one thing that's in balance by another thing that's also in balance, so the result must be in balance too!

Stu – Make sense!

Ed – OK - here goes step eight:

$$XX^m = \gamma\left(X^m + \frac{VX^m}{C}\right) \times \gamma\left(X - \frac{VX}{C}\right)$$

$$XX^m = \left(\gamma X^m + \frac{\gamma VX^m}{C}\right) \times \left(\gamma X - \frac{\gamma VX}{C}\right)$$

$$XX^m = \gamma^2 XX^m - \frac{\gamma^2 VXX^m}{C} + \frac{\gamma^2 VXX^m}{C} - \frac{\gamma^2 V^2 XX^m}{C^2}$$

$$XX^m = \gamma^2 XX^m - \frac{\gamma^2 V^2 XX^m}{C^2}$$

$$1 = \gamma^2 - \frac{\gamma^2 V^2}{C^2}$$

$$1 = \gamma^2 \left(1 - \frac{V^2}{C^2}\right)$$

$$\gamma^2 = \frac{1}{1 - \frac{V^2}{C^2}}$$

$$\gamma = \frac{1}{\sqrt{1 - \left(\frac{V}{C}\right)^2}} \qquad (8)$$

Figure 24

Stu – Wow! So the adjustment factor, gamma, equals "one over the square root of one minus 'V' over 'C' squared". That's quite a mouthful! I can see why it's called gamma for short!

Ed – Indeed!

CHAPTER 5 – TRANSFORMING TIME

Stu – So now do we have all the information we need to update the table we were making earlier for the opening of the train doors, to show, not what *intuitively* happens, but what *actually* happens?

Ed – Let's see. Let's work through the table one row at a time, and see how far we can get.

Stu – OK.

Ed – The first thing we need to do is to go back to the very first rows we did. Do you remember these from step one?

		Stu Passenger	Ed Bystander
Light Speed	The speed of light to the front door	①C	①(C + V)
	The speed of light to the back door	①−C	①(−C + V)

Figure 4

Stu – Yes. It shows different speeds for you and me for light. But we know that's not correct, so for step nine, our table would start out like this:

		Stu Passenger	Ed Bystander
Light Speed	The speed of light to the front door	⑨C	
	The speed of light to the back door	⑨−C	

Figure 25

Ed – That's right. And we've already worked out that the distance the light travels to the front door for me as bystander is:

$$X = \gamma(X^m + VT^m) \quad (6)$$

Figure 18

... so for step ten let's update our table with that:

		Stu Passenger	Ed Bystander
Light Speed	The speed of light to the front door	c	c
	The speed of light to the back door		$-c$
Light Distance	The distance the light has to travel to the front door	X^m	(10) $\gamma(X^m + VT^m)$
	The distance the light has to travel to the back door		?
Time	The time elapsed before the front door opens	$T^m = \dfrac{X^m}{c}$	?
	The time elapsed before the back door opens		?

Figure 26

Stu – So I guess that for the back door it's the same kind of thing?

Ed – Yes. That will be step eleven. Now, when we did step four to work out what we'd intuitively think it would be, we came up with:

$$-X = -X^m + VT^m \quad (4)$$

Figure 13

... and since we now know that any distance that you measure in your private universe I must multiply by gamma, then it will be:

$$-X = \gamma(-X^m + VT^m)$$

Figure 27

Stu – Yes, that makes sense.

Ed – Good, but before we update our table, let's rationalise it a bit.

Stu – How can we do that?

Ed – Well, at the moment we have an equation for each door. We have a "*plus*" version for the front door and, because it's in the opposite direction, a "*minus*" version for the back door:

$$X = \gamma(X^m + VT^m) \quad (6)$$

Figure 18

$$-X = \gamma(-X^m + VT^m)$$

Figure 27

Stu – Yes, that's because we're reversing the distances between the lantern and the front door, i.e., "X" for you and "X^m" for me, to get the distances to the back door, i.e., "-X" for you and "-X^m" for me.

Ed – That's right. But if we *started* by saying that the distance to the back door is *negative*, then we wouldn't need two equations, only one.

Stu – Really?

Ed – I can show you with an example. Let's say "X^m", the distance from the lantern to each door for you, is 3 metres as before, and "VT^m", the distance the train moves between the lantern being switched on and the light reaching the door, is 1 metre.

Stu – OK. Then the "plus" version of the formula would give the value of "X", i.e., the distance between you as bystander and the front door, to be "3 plus 1" equals 4 metres, multiplied by gamma, and the "minus" version would give "minus 3 plus 1" equals minus 2 metres, again multiplied by gamma.

Ed – That's right. But if we described the distance between the lantern and the back door as *minus* 3 metres, then the first formula would give a value for "X" of "minus 3 plus 1", which also equals minus 2 metres, multiplied by gamma.

Stu – I get it! We can use the same formula for both directions, i.e., backwards *and* forwards, as long as we take account of the direction *before* we input the numbers.

Ed – Exactly. And that applies to the light from your lantern too. So we can rationalise our table to look like this:

EINSTEIN'S RELATIVITY

		Stu Passenger	Ed Bystander
Light Speed	The speed of light	C	
Light Distance	The distance the light has to travel to a door	X^m	$X = \gamma(X^m + VT^m)$ (11)
Time	The time elapsed before a door opens	$T^m = \dfrac{X^m}{C}$	?

Figure 28

Stu – That looks much simpler. So now all we have to work out is the times at which the doors open for you?

Ed – Indeed. That will be step twelve. We can work that out from the formulas we've just derived for distance. Remember we worked out earlier that the time between the lantern being switched on, and you seeing each door open, would be the distance the light has to travel to the door, i.e., "X^m", divided by the speed of light, "C":

$$T^m = \frac{X^m}{C} \qquad (3)$$

From Figure 7a

… which means that the distance the light has to travel to each door for you, i.e., "X^m", is equal to the time between the lantern being switched on and you seeing each door open, i.e., "T^m", multiplied by the speed of light, "C", so:

$$X^m = CT^m$$

Figure 29

Stu – I get that. It's because "distance equals speed times time", like we said at the start.

Ed – Exactly. Now, if that's the distance the light has to travel to each door for you, then what do you think it will be for me?

Stu – Well, again, because "distance equals speed times time", it will be:

$$X = CT$$

Figure 30

Ed – Right. Now, if we go back to step six and the distance the light travels to the front door for me:

$$X = \gamma(X^m + VT^m) \qquad (6)$$

Figure 18

… and knowing, like we just said, that:

$$T^m = \frac{X^m}{C} \;(3)$$
From Figure 7a

$$X^m = CT^m$$
Figure 29

$$X = CT$$
Figure 30

… then we can substitute individually for "T^m", "X^m" and "X" to get this:

$$CT = \gamma\left(CT^m + \frac{VX^m}{C}\right)$$

Figure 31

Stu – So we've swapped the distances with functions of times, and vice-versa?

Ed – That's right! And if, for step twelve, we divide both sides by "C", then for the time it takes for the front door to open for me we get:

$$T = \gamma\left(T^m + \frac{VX^m}{C^2}\right) \quad \text{\textcircled{12}}$$

Figure 32

… which is the "Lorentz transformation" for time!

Stu – OK! We've already worked out the Lorentz transformation for distance, and we said that it would enable us to transform *where* things happen between our private universes. I guess this new transformation does the same thing for *when* things happen?

Ed – Exactly! If we know at what time something happens for you, then with the formula we've just derived we'll be able to work out at what time it happens for me!

Stu – So now we've got both when *and* where! And since an event is defined as something that happens at a particular time and place, then we must have all we need to transform events between our private universes!

Ed – That's right! Let's finalise out table:

		Stu Passenger	Ed Bystander
Light Speed	The speed of light	\multicolumn{2}{c}{c}
Light Distance	The distance the light has to travel to a door	X^m	$X = \gamma(X^m + VT^m)$
Time	The time elapsed before a door opens	$T^m = \dfrac{X^m}{c}$	$T \stackrel{\text{(12)}}{=} \gamma\left(T^m + \dfrac{VX^m}{c^2}\right)$

Figure 33

Stu – OK - I can see that, whereas before we had the same formula for both of us for when the doors open, we now have two formulas – one for my private universe and one for yours.

Ed – Yes. And the formula for my private universe will give two different times, one for each door, because "X^m" will be *plus* 3 metres for the front door, and *minus* 3 metres for the back door.

Stu – So in total there will be three different times, one for me, and two for you?

Ed – That's right! Before we had three different speeds for light, and a single time for both of us seeing both doors opening. Now we have a single speed for light, and consequently three different times for the opening of the doors. I see the doors open at different times, and both of those times are different to what you see!

Stu – Wow!

CHAPTER 6 – EINSTEIN MEETS PYTHAGORAS

Ed – Now, with a little change to our thought experiment we can not only confirm the formula we've worked out for gamma, but also visualise time passing at different rates in each of our private universes!

Stu – Sounds interesting! What change to our thought experiment do we need to make?

Ed – Well, currently we intuitively expect the speed of light to be increased by the speed of the train, and we can change our thought experiment so that's no longer the case.

Stu – How can we do that?

Ed – Well, at the moment the light and the train are travelling in the same direction, or to be more precise, the same dimension, i.e., backwards and forwards.

Stu – So, does that mean that the change to our thought experiment that you're talking about is to have the light travelling in a *different* dimension to the one in which the train is travelling?

Ed – Exactly. We'll start thinking about the light that shines vertically from your lantern, up to the carriage ceiling.

Stu – OK!

Ed – But before we do that we're going to take another short detour – this time into the famous theorem of Pythagoras.

Stu – What's Pythagoras got to do with relativity?

Ed – Well, do you remember the Pythagoras theorem I'm talking about?

Stu – I think it's something like, "in a right-angled triangle, the square of the longest side, equals the sum of the squares of the other two sides".

Ed – That's right. And that's going to be very useful to us because, in our second thought experiment, the light and the train are going to be travelling at right-angles to each other. And the property of right-angled triangles that you've just described is the key to quantifying the differences between our private universes. Now, since we're taking things step by step, let's not just take Pythagoras' theorem for granted, let's see why it's true.

Stu – We can do that?

Ed – Yes, it's not so difficult!

Stu – OK, I'm ready.

Ed – Good. Let's start with a simple rectangle like this:

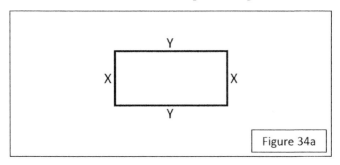

Figure 34a

... now, do you remember how to calculate the area of a rectangle?

Stu – Yes, it's width times length, so in this case, it's "X times Y".

Ed – Good. Now let's divide our rectangle into two equal parts like this:

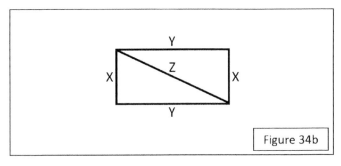

Figure 34b

Stu – Looks like you've just created two right-angled triangles!

Ed – That's right. What do you think their area will be?

Stu – Well, since you've divided the square into two equal parts, then the area of each triangle must be half the area of the rectangle, i.e., "'X times Y' divided by two", or "(X x Y) / 2".

Ed – Good. Now, let's clone our rectangle so that we have two identical ones:

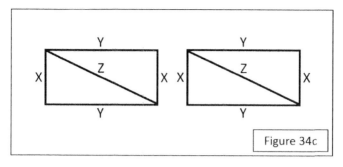

Figure 34c

… which we can separate into four identical right-angled triangles like these:

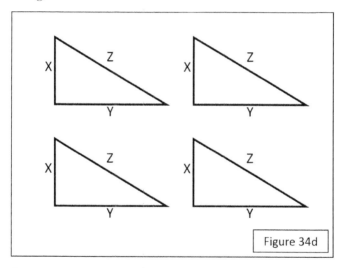

Figure 34d

Stu – Why do we need four triangles that are all the same?

Ed – Because we can arrange them to make a new rectangle like this:

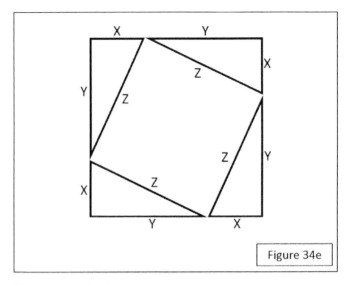

Figure 34e

Stu – Interesting!

Ed – Indeed. Over to you now. How do you think we should calculate the area of our new rectangle?

Stu – Well, it will be the length of one side multiplied by the length of the other, i.e., "X + Y" times "X + Y".

Ed – Yes, that's right - but we can also calculate it by adding up the areas of the four triangles together with the smaller rectangle in the middle – the one created by the longest sides, also known as the "hypotenuses", of the four triangles.

Stu – Yes, I see. That would be the area of each of the four triangles which is "(X x Y) / 2", plus the area of the smaller rectangle in the middle, which is "Z times Z", or "Z^2". So the total area of our new rectangle will be:

$$\text{Area} = \left(4 \times \frac{X \times Y}{2}\right) + Z^2$$

$$\text{Area} = \left(\frac{4}{2} \times XY\right) + Z^2$$

$$\text{Area} = 2XY + Z^2$$

Figure 35

Ed – Good. Now we've got two different ways of working out the area of our new rectangle – the first one that you came up with, i.e., "(X + Y) times (X + Y)", and the one you've just worked out of "2XY + Z²".

Stu – And since they're two different ways of calculating the same thing, that must mean that "(X + Y) times (X + Y)" equals "2XY + Z²"!

Ed – Indeed! Next I'm going to work out what "(X + Y) times (X + Y)" equals. To do that I have to multiply the "X" and the "Y" in the first bracket by the "X" and the "Y" in the second one like this:

$$\text{Area} = (X + Y) \times (X + Y)$$

$$\text{Area} = XX + XY + YX + YY$$

$$\text{Area} = X^2 + XY + XY + Y^2$$

> **Area = $X^2 + 2XY + Y^2$**
>
> Figure 36

Stu – OK, so now, instead of saying that "$2XY + Z^2$" equals "$(X + Y)$ times $(X + Y)$", we can say that it equals "$X^2 + 2XY + Z^2$".

Ed – That's right. Do you think you can take it from here? I mean, to put our two formulas together, and see what comes out?

Stu – OK. Here we go:

> $X^2 + 2XY + Y^2 = 2XY + Z^2$
>
> $X^2 + 2XY + Y^2 - 2XY = Z^2$
>
> $X^2 + Y^2 = Z^2$
>
> Figure 37

Ed – Good! Do you see what you've just done?

Stu – Well, it seems like I've just proved that the square of the hypotenuse equals the sum of the squares of the other two sides - just like Pythagoras said!

Ed – Indeed! And like we said earlier, that's going to come in very useful!

CHAPTER 7 – TIME DILATION

Ed – So let's go back to our train and lantern again, and our second thought experiment.

Stu – Oh yes, I'd forgotten about that!

Ed – This time, let's imagine that your lantern is on the floor of the carriage. Also let's imagine a trap door controlled by a photo-electric cell in the ceiling of the carriage directly above the lantern. And we'll call the distance between the lantern and the trap door, i.e., the height of the carriage, "H^m". Same as before, the little "m" means that it's not in my private universe, but in yours which is moving relative to mine. Let's draw a diagram to illustrate the light travelling up to the trap door:

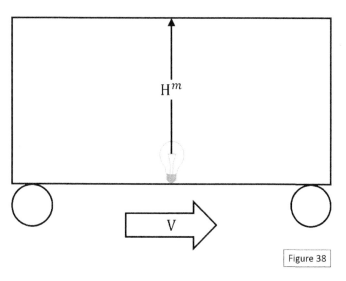

Figure 38

Stu – OK, so as we said, unlike before, the light and the train are now travelling in different dimensions – the train is travelling forwards, or horizontally, while the light is travelling upwards, or vertically.

Ed – That's right. Next, let's think about the distance the light has to travel up to the trap door for me in my private universe. We'll call that distance "H", without a little "*m*". What do we know about it?

Stu – Well, just like in our first thought experiment, the train will have travelled some distance between me switching on the lantern and the trap door opening, so the distance the light has to travel for you will be different, because the trap door will have moved with the train.

Ed – Indeed. Let's update our diagram to illustrate that:

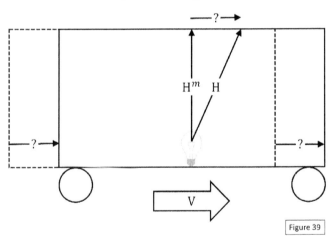

Figure 39

Stu – OK, same as before, the light has to travel further for you in your private universe.

Ed – Yes. Now the next thing we need to do, also just like before, is to work out the distance that the train travels between you switching on the lantern and the trap door opening. Let's consider what that distance is for me in my private universe as bystander - do you think you can work it out?

Stu – Let me try. We know that "distance equals speed times time". The speed of the train is "V", but we haven't yet worked out the time it takes for the light to reach the trap door for you.

Ed – That's right. So, for now, let's call that time for me as bystander "T".

Stu – OK. So the distance the train travels between the lantern being switched on and the trap door opening would be "V" times "T", or "VT".

Ed – Yes. Let's add that to our diagram:

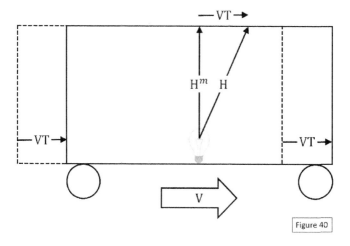

Figure 40

Stu – Looks like we've just created a right-angled triangle, with a hypotenuse of length "H", and the other two sides of length "H^m" and "VT". And because it's a right-angled triangle, we know that "$H^2 = H^{m2} + (VT)^2$".

Ed – Precisely!

Stu – But how will that help us calculate the rate at which time passes in each of our private universes? The formula of "$H^2 = H^{m2} + (VT)^2$" doesn't include the time it takes for the light to reach the trap door for either of us – it only has the distances!

Ed – Indeed, but we can easily amend it so that it has the times.

Stu – How can we do that?

Ed – Well, let's take "H", the distance the light has to travel for me in my private universe to reach the trap door. What do we know about that distance?

Stu – Well, all I can think of is that it must equal "speed times time".

Ed – That's right. And that's all we need to know. What do you think the speed will be?

Stu – As we know, the speed of light is always "C".

Ed – Indeed. And what do you think the time will be?

Stu – Well, we already said that we'd call the time it takes for the light to reach the trap door for you "T".

Ed – Yes, so the distance will be "C" times "T", or "CT".

Stu – Makes sense.

Ed – Indeed. Now what about "H^m"?

Stu – Well, again the speed of light is "C", and if we call the time it takes for the light to reach the trap door for me in my private universe "T^m", then "H^m" will equal "C times T^m", or "CT^m".

Ed – Right. Now we can go back to our formula of "$H^2 = H^{m2} + (VT)^2$" and replace "H" with "CT", and "H^m" with, "CT^m". Then our formula will be "$(CT)^2 = (CT^m)^2 + (VT)^2$".

Stu – OK, so now we've got a formula with everything expressed in terms of speed and time – no distances anymore!

Ed – That's right! Now we know, not only the length of each side of the triangle expressed in terms of speed and time, but thanks to Pythagoras, we also know the relationship between them. Let's update our diagram and do the same thing that we just did in our formula, i.e., replace "H" with "CT", and "H^m" with "CT^m":

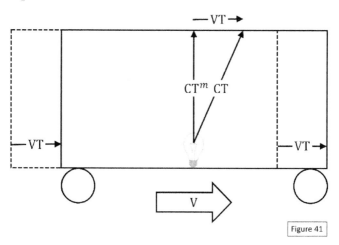

Figure 41

Stu – So now we can calculate the different rate at which time passes in our two private universes?

Ed – That's right! So let's do it:

$$(CT)^2 = (CT^m)^2 + (VT)^2$$

$$(CT^m)^2 = (CT)^2 - (VT)^2$$

$$C^2T^{m2} = C^2T^2 - V^2T^2$$

$$T^{m2} = \frac{C^2T^2 - V^2T^2}{C^2}$$

$$T^{m2} = \frac{T^2(C^2 - V^2)}{C^2}$$

$$T^{m2} = T^2\left(\frac{C^2}{C^2} - \frac{V^2}{C^2}\right)$$

$$T^{m2} = T^2\left(1 - \frac{V^2}{C^2}\right)$$

$$T^m = T\sqrt{1 - \left(\frac{V}{C}\right)^2}$$

$$T = T^m \times \frac{1}{\sqrt{1 - \left(\frac{V}{C}\right)^2}}$$

Figure 42

… which is known as the "Lorentz factor" for time. Do you recognise anything about it, Stu?

Stu – Yes! "One over the square root of one minus 'V' over 'C' squared", i.e., the formula for gamma that we worked out in step eight in our first thought experiment!

Ed – Indeed! So having spotted gamma, for step thirteen we can write the Lorentz factor for time like this:

$$T = \gamma T^m \qquad \text{}$$

Figure 43

Stu – OK! Can we use our new Lorentz factor in an example?

Ed – Sure. Let's assume that in your private universe on the train the distance the light travels from the lantern to the trap door in the ceiling is 3 metres.

Stu – OK. Since that's the same distance as the distance from the lantern to the back and front doors, then we know from our first thought experiment that for me on the train it will take 10 nanoseconds for the trap door in the ceiling to open, so that's what "T^m" will be.

Ed – That's right. Now let's use the Lorentz factor to work out what "T" will be, i.e., how much time will pass for me during the 10 nanoseconds that pass for you. Let's assume that the train is travelling at 180,000 kilometres per second, i.e., 60% of the speed of light, or 0.6C, and let's first work out the value of gamma using the formula we worked out in step eight:

EINSTEIN'S RELATIVITY

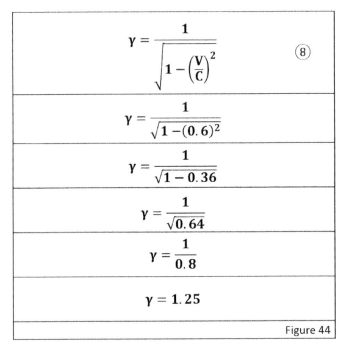

$$\gamma = \frac{1}{\sqrt{1-\left(\frac{V}{C}\right)^2}} \quad \text{⑧}$$

$$\gamma = \frac{1}{\sqrt{1-(0.6)^2}}$$

$$\gamma = \frac{1}{\sqrt{1-0.36}}$$

$$\gamma = \frac{1}{\sqrt{0.64}}$$

$$\gamma = \frac{1}{0.8}$$

$$\gamma = 1.25$$

Figure 44

… and remembering what we just worked out in step thirteen:

$$T = \gamma T^m \quad \text{⑬}$$

Figure 43

… then:

$T = \gamma T^m$ (13)
$T = 1.25 \times 10$
$T = 12.5$ nanoseconds

Figure 45

Stu – So 12.5 nanoseconds will have passed before the trap door opens for you, while only 10 nanoseconds will have passed for me – so we've shown that time slows down for moving objects!

Ed – Precisely! For every 12.5 nanoseconds that go by on my watch, only 10 will go by on yours. Or instead of nanoseconds it could be seconds, minutes, hours, days, months, etc., so you'll age slower than me!

Stu – Wow! So if you'll see me age slower than you, then does that mean that, when I look into your private universe, I'll see you age faster than me?

Ed – That's what you might intuitively expect, but once again, that's not what actually happens!

Stu – Really! So what does happen then?

Ed – Well, do you remember what we said earlier about relativity being symmetrical, and what you experience being the mirror-image of what I experience?

Stu – Yes, I remember that.

Ed – So what actually happens is that you'll see me age slower than you, just like I'll see you age slower than me!

Stu – Now that really is counter-intuitive!

Ed – Indeed – the consequences of relativity take some getting used to!

Stu – They certainly do! Could you explain how the Lorentz *factor* for time that we've just worked out in our second thought experiment differs from the Lorentz *transformation* for time that we worked out in the first one?

Ed – Sure. Let's put the two formulas side by side so we can compare them – the factor on the left and the transformation on the right:

$$T = \gamma T^m \quad (13)$$

Figure 43

$$T = \gamma \left(T^m + \frac{VX^m}{C^2} \right) (12)$$

Figure 32

Stu – They look the same except the transformation includes "VX^m over C^2", which must be some measure of distance because it's the distance to a door in my private universe, multiplied by the speed of the train, all divided by the speed of light squared.

Ed – That's right. The Lorentz *factor* for time is relevant to time that passes irrespective of distance, i.e., in the same place, like for example the ticks of the watch on your wrist. Whereas the Lorentz *transformation* for time also takes into account the fact that things may not have happened in the same place – that's the reason for the "VX^m over C^2". For example, when we think about our first thought experiment, the front and back doors of your carriage are in different places. That's why I see them opening at different times. Or to put it another way, if the doors were in the same place, then I'd see them open at the same time.

Stu – That reminds me of the "shifting" of time that we mentioned earlier, when we talked about the times at which we each experience events being shifted backwards and forwards. We said that, because the speed of light isn't shifted, i.e., sped-up, by the speed of the train, then the times at which the doors open are shifted instead.

Ed – Exactly! Motion causes time to both dilate *and* to shift. The Lorentz factor only takes account of the dilation, whereas the Lorentz transformation takes account of both the dilation *and* the shifting.

Stu – Got it!

Ed – OK. Now, if we look at the time transformation formula again:

$$T = \gamma \left(T^m + \frac{VX^m}{C^2} \right) \quad \text{(12)}$$

Figure 32

… it's telling us that the time it takes for *me* for the light to travel to the front door is equal to the time it takes for you, *plus* the time shift, all multiplied by gamma. So, because relativity is symmetrical, that means that the time it takes for *you* for the light to travel to the front door is equal to the time it takes for me, *minus* the time shift, all multiplied by gamma. So for step fourteen:

EINSTEIN'S RELATIVITY

$$T^m = \gamma \left(T - \frac{VX}{C^2} \right) \quad (14)$$

Figure 46

Stu – Makes sense. It's like what we did earlier to create the mirror image of the transformation for distance.

Ed – That's right. So now we know how to transform time from your private universe to mine, and vice-versa!

CHAPTER 8 – LENGTH CONTRACTION

Stu – Our second thought experiment has already shown us how time slows down for moving objects. Can it tell us anything about what happens to distance? It looks to me like distance has increased for you, because the light has to travel further to reach the trap door in the ceiling.

Ed – Yes, but don't forget that the train has moved in between the lantern being switched on and the light reaching the trap door!

Stu – Yes, that's right! And also, distance shrinks for moving objects – not increases!

Ed – Quite right. Let's figure it out this way. Imagine the train is moving away from me as bystander, and I lay a stick down parallel to the track. The stick is therefore in my private universe. For me as bystander let's call the distance between the two ends of the stick "X", i.e., its length. For you as passenger the length of the stick will be "X^m", because it's not in your private universe.

Stu – OK. Like we said earlier, motion is relative, so the stick is moving relative to me, just like the train is moving relative to you.

Ed – That's right. Now, for *me*, how long will it take you in your private universe to pass from one end of the stick to the other, i.e., what will "T^m" be?

Stu – Well, we know that "time equals distance divided by speed" so it will be the length of the stick for you, "X" divided by the speed of the train "V", so "X over V", i.e:

$$T^m = \frac{X}{V}$$

Figure 47

Ed – Good. And we can rearrange that to calculate the length of the stick for me as:

$$X = VT^m$$

Figure 48

… now, how long will it take you in your private universe on the train to pass the stick, i.e., what will "T" be?

Stu – Well, like we said, since the stick isn't in my private universe it's length will be "X^m", and since "time equals distance divided by speed" then the time it will take me to pass the stick will be the length of the stick, "X^m", divided by the speed of the train "V", so "X^m over V", i.e:

$$T = \frac{X^m}{V}$$

Figure 49

Ed – Good. And again, we can rearrange that to calculate the length of the stick for you as:

$$X^m = VT$$

Figure 50

Stu – OK so far!

Ed – Good. Now, so far we've got:

$$X = VT^m$$

Figure 48

$$X^m = VT$$

Figure 50

… so the ratio between "X^m" and "X" will be:

$$\frac{X^m}{X} = \frac{VT}{VT^m}$$

Figure 51

… and because the "V's" on the top and bottom of the right-hand side of the equation cancel each other out, then:

$$\frac{X^m}{X} = \frac{T}{T^m}$$

Figure 52

Now, we worked out earlier that:

$$T = \gamma T^m \qquad \text{(13)}$$

Figure 43

... which we can rearrange as:

$$\gamma = \frac{T}{T^m}$$

Figure 53

... so that means that, for step fifteen:

$$\frac{X^m}{X} = \gamma$$

$$X = \frac{X^m}{\gamma} \quad (15)$$

Figure 54

... which is the Lorentz factor for distance!

Stu – OK! So the adjustment factor for distance is the inverse, or reciprocal, of the adjustment factor for time?

Ed – That's right! Time slows down, i.e., the interval between seconds gets longer, whereas distance shrinks, i.e., gets shorter.

Stu – Makes sense – the stick is shorter for me because, due to time dilation, it takes me less time to pass it, so for me it's shrunk. Can we work out an example for distance like we did for time?

Ed – Sure. Let's again assume that the train is travelling at 0.6C. We already know from our time dilation example that, at that speed, gamma equals 1.25. We also know that for you the distance from the lantern to the front door, i.e., "X^m", is 3 metres, so the distance for me will be:

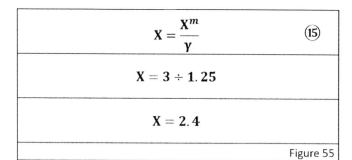

Figure 55

Stu – So the distance to the front door will only be 2.4 metres for you?

Ed – Yes! Moving objects shrink, like we just said!

Stu – And what about the distance to the trap door in the ceiling?

Ed – Good question - that will be 3 metres for me, just like it is for you!

Stu – Really? Why's that?

Ed – It's because objects only shrink in the dimension in which they're moving – so our train will get less long, but not less tall, or less wide.

Stu – OK! Could you spell out the difference between the Lorentz *factor* for distance and the Lorentz *transformation* for distance, like you did earlier for the time factor and transformation?

Ed – Sure. Let's again put them side by side – the factor on the left, and the transformation on the right:

$X = \dfrac{X^m}{\gamma}$ ⑮
From Figure 54

$X = \gamma(X^m + VT^m)$ ⑥
Figure 18

Stu – That looks strange! For the factor we're *dividing* distance in my private universe by gamma to get distance in yours, but the transformation includes *multiplying* distance in my private universe by gamma to get distance in yours!

Ed – Yes, but as we can see when we look at the two formulas, the factor takes no account of time, but the transformation does. Let's remember something we did earlier in step seven:

$X^m = \gamma(X - VT)$ ⑦
Figure 21

Stu – Yes, I remember that. It was just before we worked out the formula for gamma in our first thought experiment. We said that the distance the light travels to the front door for me is equal to the distance the light travels to the front door for you, minus the distance I've travelled on the train between the lantern being switched on and the light reaching the door, all multiplied by gamma.

Ed – That's right. And do you also remember what we said about measuring the length of a moving object?

Stu – Yes. We said that to measure the length of a moving object, you have to measure both ends at the same time, otherwise it will have moved, and you'll get the wrong measurement.

Ed – That's right. So if for me both ends of an object, for example your carriage, are to be measured at the same time, i.e., if there's to be no movement for me between the two measurements, then the "VT" part of the equation must be zero, i.e., no time can pass for me between the measurements, and no distance can be travelled by you.

Stu – OK. So if no time passes for you, that means "T" must be equal to zero, so:

$$X^m = \gamma[X - (V \times 0)]$$

Figure 56

Ed – Yes. And since "V" times zero is zero, then:

$$X^m = \gamma X$$

Figure 57

….so:

$$X = \frac{X^m}{\gamma}$$

Figure 58

… which is the Lorentz factor from step fifteen!

Stu – That's right! So the factor gives the value for "X", i.e., distance in your private universe, when there *isn't* any time difference and movement between measurements, and the transformation gives the value for "X" when there *is*?

Ed – Precisely! So, in our first thought experiment, for me the distance between where I'm standing on the platform and the front door will have increased during the time between you turning on the lantern and the light reaching the door, because the front door will have moved away from me.

Stu – Got it! Also, it looks like we've just derived the factor formula from the transformation formula!

Ed – That's right. And since the transformation formula was derived from our first thought experiment, and the factor formula from the second, then it shows that the results of our two thought experiments are in agreement!

Stu – That's reassuring!

Ed – Indeed! And by deriving the factor from the transformation we can see that the Lorentz factor for distance can be thought of as a kind of "special case" of the transformation – special because there's no time difference.

Stu – So, if that's the case for *distance*, then does that mean that the Lorentz factor for *time* can also be thought of special case of the transformation, because there's no *distance* difference?

Ed – Precisely!

CHAPTER 9 – PUTTING IT ALL TOGETHER

Ed – Now, it helps to think about the Lorentz transformations in terms of "events". So if we say for example that "event one" is the turning on of the lantern, and "event two" is the opening of the front door, then if we know where and when those events happen for you, i.e., their "coordinates" in space and time, then like we said earlier, we can use the transformation formulas to work out where and when they happen for me.

Stu – But a distance, or a length, like for example 3 metres, isn't an event, and nor is a time period, like 10 nanoseconds!

Ed – That's true. But we can think of a distance as the difference between two events in different places at the same time, and a time period as the difference between two events at different times in the same place!

Stu – Yes, I can see that!

Ed – OK. Now, let's return to our first thought experiment with the train doors, and let's start thinking about what's going on in terms of "events" and their "co-ordinates". Would you like to try that?

Stu – OK. Let's say that I switch the lantern on at noon, and like we already worked out, I see both doors opening 3 metres away from me at 10 nanoseconds after noon. Does that give us the kind of "event coordinates" we're talking about?

Ed – Yes, that gives us exactly the kind of event coordinates we're talking about!

Stu – OK. So can we use the Lorentz transformation formulas to figure out where and when those events happen for you?

Ed – Sure we can! First let's summarise what we already know in a table, starting with what you just said. Let's give noon a time of "zero" as our starting point, i.e., the time when the first event of you switching on the lantern happens:

Coordinate		Event	
		Lantern on	Door opens
Front Door	Place	3 m	3 m
	Time	0 ns	10 ns
Back Door	Place	-3 m	-3 m
	Time	0 ns	10 ns

Figure 59

Stu – OK. When I turn on the lantern the doors are 3 metres away from me. And when they open 10 nanoseconds later, they are still 3 metres away from me, because I'm moving with them.

Ed – That's right.

Stu – And we already worked out earlier using the Lorentz factor for time that, when the train is travelling at 60% of the speed of light, you'd see 12.5 nanoseconds pass on your watch for the 10 nanoseconds that pass on mine.

Ed – That's right. Let's add that to our table:

Coordinate		Event			
		Stu / Passenger		Ed / Bystander	
		Lantern on	Door opens	Lantern on	Door opens
Front Door	Time	0 ns	10 ns	?	?
	Place	3 m	3 m	?	?
Lantern & Stu	Time	0 ns	10 ns	?	12.5 ns
	Place	0 m	0 m	?	?
Back Door	Time	0 ns	10 ns	?	?
	Place	-3 m	-3 m	?	?

Figure 60

Stu – I see you've added another location to our table, i.e., where I am with the lantern.

Ed – Yes, we need to do that because it's the location of your watch. It's the place you were when you saw the 10 nanoseconds pass between switching on the lantern and the doors opening. It's to do with what we said earlier about time dilation being for events that happen at the same place, so that there won't have been any "shifting" of time due to distance.

Stu – OK. And I guess we can already fill in several of those question marks?

Ed – Indeed we can. We know that you and I were in the same place when you switched on the lantern, because you switched it on at the exact moment that you passed me. And we also know that it was noon for both of us when you did that. So if we update our table with that information then it will look like this:

Coordinate		Event			
		Stu / Passenger		Ed / Bystander	
		Lantern on	Door opens	Lantern on	Door opens
Front Door	Time	0 ns	10 ns	0 ns	?
	Place	3 m	3 m	?	?
Lantern & Stu	Time	0 ns	10 ns	0 ns	12.5 ns
	Place	0 m	0 m	0 m	?
Back Door	Time	0 ns	10 ns	0 ns	?
	Place	-3 m	-3 m	?	?

Figure 61

Stu – OK. And I remember that we also worked out what the length contraction would be.

Ed – Yes. We worked out that the front door will be 2.4 metres from the lantern to me.

Stu – So then the distance to the back door would shrink in the same way?

Ed – That's right. Let's update our table:

Coordinate		Event			
		Stu / Passenger		Ed / Bystander	
		Lantern on	Door opens	Lantern on	Door opens
Front Door	Time	0 ns	10 ns	0 ns	?
	Place	3 m	3 m	2.4 m	?
Lantern & Stu	Time	0 ns	10 ns	0 ns	12.5 ns
	Place	0 m	0 m	0 m	?
Back Door	Time	0 ns	10 ns	0 ns	?
	Place	-3 m	-3 m	-2.4 m	?

Figure 62

Stu – I see you've put those distances in the column for the event of me switching on the lantern.

Ed – Yes. That's because of what we said earlier about length contraction being about distances that are measured at the same time. We can't measure the length of the train for me as being the difference between the distance to the front door when it opens and the distance to the back door when it opens, because they open at different times!

Stu – I see.

Ed – Now, so we can focus on the differences between our private universes, let's replace all the zero values with a "dash" like this, "-". So that will be all coordinates concerning the switching on of the lantern, which happens at noon for both of us, and also in the same place, i.e., where you pass me. Then our table will look like this:

Coordinate		Event			
		Stu / Passenger		Ed / Bystander	
		Lantern on	Door opens	Lantern on	Door opens
Front Door	Time	-	10 ns	-	?
	Place	3 m	3 m	2.4 m	?
Lantern & Stu	Time	-	10 ns	-	12.5 ns
	Place	-	-	-	?
Back Door	Time	-	10 ns	-	?
	Place	-3 m	-3 m	-2.4 m	?

Figure 63

Stu – Yes, that helps highlight the differences. So now can we work out the values for the rest of the question marks?

Ed – Indeed we can! Let's start with you and the lantern. We've already worked out the time coordinate for that location for both of us, so let's use the Lorentz transformation to work out the other coordinate, i.e., distance. Here's the formula we derived:

$$X = \gamma(X^m + VT^m) \qquad (6)$$

Figure 18

Stu – We've already worked out that 60% of the speed of light is 180,000 kilometres per second, and at that speed gamma equals 1.25. So we already know "V" and "γ".

Ed – That's right. Now, "X^m", i.e., the distance between you and the lantern in your private universe is zero, because you and the lantern are in the same place. And "T'''", i.e., the time between you switching on the lantern and the front door opening, is 10 nanoseconds, so:

$$X = \gamma(X^m + VT^m) \quad (6)$$

$$X = 1.25 \times [0 + (180,000 \times 10 \text{ ns})]$$

$$X = 1.25 \times (180,000 \times 10 \text{ ns})$$

$$X = 1.25 \times \left(180,000 \times \frac{10}{1,000,000,000}\right)$$

$$X = 1.25 \times \left(\frac{1,800,000}{1,000,000,000}\right)$$

$$X = 1.25 \times 0.0018 \text{ kilometres}$$

$$X = 1.25 \times 1.8 \text{ metres}$$

$$X = 2.25 \text{ metres}$$

Figure 64

Stu – So, for you the lantern and I would move 2.25 metres in the time between me switching it on and the light reaching the front door?

Ed – That's right. Let's include that result in our table:

Coordinate		Event			
		Stu / Passenger		Ed / Bystander	
		Lantern on	Door opens	Lantern on	Door opens
Front Door	Time	-	10 ns	-	?
	Place	3 m	3 m	2.4 m	?
Lantern & Stu	Time	-	10 ns	-	12.5 ns
	Place	-	-	-	2.25 m
Back Door	Time	-	10 ns	-	?
	Place	-3 m	-3 m	-2.4 m	?

Figure 65

Stu – OK. Now can we work out the locations of the doors for you when they open?

Ed – Yes, let's do that next. We already know from our last calculation that "VT^m" equals 1.8 metres – that's how far the train travels for you in 10 nanoseconds at 180,000 kilometres per second. And we know that for you the distance between the lantern and both doors is 3 metres, so:

Back Door	Front Door
$X = \gamma(X^m + VT^m)$ ⑥	
$X = 1.25 \times (-3 + 1.8)$	$X = 1.25 \times (3 + 1.8)$

$X = 1.25 \times -1.2$	$X = 1.25 \times 4.8$
$X = -1.5$ metres	$X = 6$ metres

Figure 66

… and updating our table with those results, then:

Coordinate		Event			
		Stu / Passenger		Ed / Bystander	
		Lantern on	Door opens	Lantern on	Door opens
Front Door	Time	-	10 ns	-	?
	Place	3 m	3 m	2.4 m	6 m
Lantern & Stu	Time	-	10 ns	-	12.5 ns
	Place	-	-	-	2.25 m
Back Door	Time	-	10 ns	-	?
	Place	-3 m	-3 m	-2.4 m	-1.5 m

Figure 67

Stu – OK! So all that's left is to work out at what times you see the doors open!

Ed – Indeed! Earlier we worked out the Lorentz transformation for time as:

$$T = \gamma \left(T^m + \frac{VX^m}{C^2} \right) \quad \text{(12)}$$

Figure 32

... now, we know that "T^m" equals 10 nanoseconds, and the distance to the front door for you is 3 metres, or 0.003 kilometres, so:

$$T = \gamma\left(T^m + \frac{VX^m}{C^2}\right) \quad \text{(12)}$$

$$T = 1.25 \times \left[\frac{10}{1,000,000,000} + \frac{180,000 \times 0.003}{300,000^2}\right]$$

$$T = 1.25 \times \left[0.00000001 + \left(\frac{540}{90,000,000,000}\right)\right]$$

$$T = 1.25 \times (0.00000001 + 0.000000006)$$

$$T = 1.25 \times 0.000000016$$

$$T = 0.00000002 \text{ seconds}$$

$$T = 20 \text{ nanoseconds}$$

Figure 68

... and for the back door it will be the same, but with a distance of *minus* 3 metres, or *minus* 0.003 kilometres, for you, so:

$$T = \gamma\left(T^m + \frac{VX^m}{C^2}\right) \quad \text{(12)}$$

$$T = 1.25 \times \left[\frac{10}{1,000,000,000} + \frac{180,000 \times -0.003}{300,000^2}\right]$$

PRIVATE UNIVERSES

$$T = 1.25 \times \left[0.00000001 + \left(\frac{-540}{90,000,000,000}\right)\right]$$

$$T = 1.25 \times (0.00000001 - 0.000000006)$$

$$T = 1.25 \times 0.000000004$$

$$T = 0.000000005 \text{ seconds}$$

$$T = 5 \text{ nanoseconds}$$

Figure 69

... so now we can complete our table:

Coordinate		Event			
		Stu / Passenger		Ed / Bystander	
		Lantern on	Door opens	Lantern on	Door opens
Front Door	Time	-	10 ns	-	20 ns
	Place	3 m	3 m	2.4 m	6 m
Lantern & Stu	Time	-	10 ns	-	12.5 ns
	Place	-	-	-	2.25 m
Back Door	Time	-	10 ns	-	5 ns
	Place	-3 m	-3 m	-2.4 m	-1.5 m

Figure 70

Stu – OK! We've worked out all the times and places for both our private universes! Could we add a column for differences, to help us compare what happens in your private universe with what happens in mine?

Ed – Good idea! Let's do that:

Coordinate		Event					
		Stu / Passenger		Ed / Bystander		Difference	
		Lantern on	Door opens	Lantern on	Door opens	Lantern on	Door opens
Front Door	Time	-	10 ns	-	20 ns	-	10 ns
	Place	3 m	3 m	2.4 m	6 m	0.6 m	3 m
Lantern & Stu	Time	-	10 ns	-	12.5 ns	-	2.5 ns
	Place	-	-	-	2.25 m	-	2.25 m
Back Door	Time	-	10 ns	-	5 ns	-	-5 ns
	Place	-3 m	-3 m	-2.4 m	-1.5 m	0.6 m	1.5 m

Figure 71

Stu – So if we look at the differences in times, we can see that the front door opens 10 nanoseconds later for you than for me, and the back door opens 5 nanoseconds earlier.

Ed – That's right. As we said earlier, contrary to what we'd intuitively expect, i.e., *three* different speeds for light, and a *single* time for both of us seeing both doors open, there's actually a *single* speed for light, and so *three* different times for seeing the doors open. I see the doors open at different times, and both of those times are different to what you see.

Stu – Yes, one earlier, and one later.

Ed – That's right. Now, to highlight what's going on with distance, let's consider the differences between the front and back doors. So let's add a row to our table for that:

Coordinate		Event					
		Stu / Passenger		Ed / Bystander		Difference	
		Lantern on	Door opens	Lantern on	Door opens	Lantern on	Door opens
Front Door	Time	-	10 ns	-	20 ns	-	10 ns
	Place	3 m	3 m	2.4 m	6 m	0.6 m	3 m
Lantern & Stu	Time	-	10 ns	-	12.5 ns	-	2.5 ns
	Place	-	-	-	2.25 m	-	2.25 m
Back Door	Time	-	10 ns	-	5 ns	-	-5 ns
	Place	-3 m	-3 m	-2.4 m	-1.5 m	0.6 m	1.5 m
Difference between Doors	Time	-	-	-	15 ns	-	15 ns
	Place	6 m	6 m	4.8 m	7.5 m	-1.2 m	1.5m

Figure 72

Stu – There's something I find strange about that!

Ed – I guess it's the 7.5 metres as the distance between the doors for me as Bystander?

Stu – That's right! It seems the distance between the doors for you has increased by 2.7 metres in the time between me switching on the lantern and the doors opening, i.e., from 4.8 metres to 7.5 metres!

Ed – Yes, and that's to do with what we said earlier about measuring things that are moving, i.e., that you have to measure both ends at the same time. And we know that the doors don't open at the same time for me – as we can see in the new row in our table, there's a difference of 15 nanoseconds.

Stu – So, we have to take into account the fact that the train has moved during the 15 nanoseconds between the doors opening?

Ed – Exactly! Let's work how far the train will have moved during those 15 nanoseconds:

Distance = Speed × Time
Distance = $180,000 \times \dfrac{15}{1,000,000,000}$
Distance = $\dfrac{2,700,000}{1,000,000,000}$
Distance = 0.0027 kilometres
Distance = 2.7 metres
Figure 73

Stu – That explains it!

Ed – Yes, our results reconcile! Let's do a few more reconciliations to make sure that everything stacks up. Let's start by working out, according to our results, what speed I see you travelling at.

Stu – Well, for you the lantern and I move 2.25 metres in 12.5 nanoseconds, so:

$$\text{Speed} = \frac{\text{Distance}}{\text{Time}}$$
$$\text{Speed} = \frac{2.25}{1,000} \div \frac{12.5}{1,000,000,000}$$
$$\text{Speed} = \frac{2.25}{1,000} \times \frac{1,000,000,000}{12.5}$$
$$\text{Speed} = \frac{2,250,000,000}{12,500}$$
$$\text{Speed} = 180,000 \text{ km per second}$$

Figure 74

Ed – That's the result we wanted! You experience yourself to be travelling 1.8 metres in 10 nanoseconds, and I experience you to be travelling 2.25 metres in 12.5 nanoseconds. But we both experience you to be travelling at the same speed of 180,000 kilometres per second!

Stu – OK! What about the speed of light to the back door for you?

Ed – Well, for me the light to the back door travels 1.5 metres in 5 nanoseconds. So the speed would be:

$$\text{Speed} = \frac{\text{Distance}}{\text{Time}}$$
$$\text{Speed} = \frac{1.5}{1,000} \div \frac{5}{1,000,000,000}$$
$$\text{Speed} = \frac{1.5}{1,000} \times \frac{1,000,000,000}{5}$$

$$\text{Speed} = \frac{1,500,000,000}{5,000}$$

$$\text{Speed} = 300,000 \text{ km per second}$$

Figure 75

... again, that's the result we wanted! Now what about the front door?

Stu – For the front door for you the light travels 6 metres in 20 nanoseconds. So the speed would be:

$$\text{Speed} = \frac{\text{Distance}}{\text{Time}}$$

$$\text{Speed} = \frac{6}{1,000} \div \frac{20}{1,000,000,000}$$

$$\text{Speed} = \frac{6}{1,000} \times \frac{1,000,000,000}{20}$$

$$\text{Speed} = \frac{6,000,000,000}{20,000}$$

$$\text{Speed} = 300,000 \text{ km per second}$$

Figure 76

Ed – That's also the result we wanted – everything reconciles!

Stu – Yes, it all fits together! I notice that the time differences between what you and I experience are very small – just a few nanoseconds!

Ed – Yes, but we're only talking about small distances. When you start considering greater distances then the differences between what people experience in their different private universes become much more significant, as we'll now see, by sending you to Mars!

CHAPTER 10 – MARS IS CLOSER THAN YOU THINK

Stu – You're sending me to Mars!

Ed – Yes, but only in a thought experiment! Let's imagine that your railway line goes all the way to Mars.

Stu – OK!

Ed – Now, the distance to Mars is about 225 million kilometres. So if we imagine that your train is travelling even faster this time, at 90% of the speed of light, then how long would you intuitively expect it would take you to get there?

Stu – Well, 90% of the speed of light is 0.9 times 300,000 equals 270,000 kilometres per second, so:

$$\text{Time} = \frac{\text{Distance}}{\text{Speed}}$$
$$\text{Time} = 225,000,000 \div 270,000$$
$$\text{Time} = 833.33 \text{ seconds}$$
$$\text{Time} = 13.9 \text{ minutes}$$
Figure 77

Ed – OK. Now, we know that you'll experience distance and time differently to me.

Stu – That's right – distance shrinks with speed! The railway line isn't in my private universe - just like the stick lying beside the railway track that we talked about earlier. So the distance will be less!

Ed – That's right!

Stu – OK! Let's work out the value for gamma at 90% of the speed of light:

$\gamma = \dfrac{1}{\sqrt{1-\left(\dfrac{V}{C}\right)^2}}$ ⑧
$\gamma = \dfrac{1}{\sqrt{1-(0.9)^2}}$
$\gamma = \dfrac{1}{\sqrt{1-0.81}}$
$\gamma = \dfrac{1}{\sqrt{0.19}}$
$\gamma = \dfrac{1}{0.4359}$
$\gamma = 2.294$
Figure 78

... and the formula we worked out earlier for length contraction is:

$$X = \frac{X^m}{\gamma} \quad \text{\textcircled{15}}$$

From Figure 54

… so for me the length of my railway line to Mars will be:

$$X = \frac{X^m}{\gamma} \quad \text{\textcircled{15}}$$

$$X = \frac{225,000,000}{2.294}$$

$$X = 98,081,953 \text{ kilometres}$$

Figure 79

Ed – Correct!

Stu – So that means it won't take me 13.9 minutes, it will take me:

$$\text{Time} = \frac{\text{Distance}}{\text{Speed}}$$

$$\text{Time} = 98,081,953 \div 270,000$$

$$\text{Time} = 363.27 \text{ seconds}$$

$$\text{Time} = 6.1 \text{ minutes}$$

Figure 80

Ed – Indeed. And we could have got the same result by

using the time dilation formula, and dividing 13.9 minutes by gamma.

Stu – OK! So by travelling very fast, not only do I get to where I'm going sooner because of my high speed, but even more so, because I've got less distance to travel – it's win-win!

Ed – That's right! And as you can see, this time the differences in time and distance are much larger than in our previous two thought experiments. The faster you go, the larger gamma becomes, and the further you go, the differences in distance and time increase too! If you travelled far enough, like beyond the Milky Way, you could end up being years younger than me!

Stu – Amazing!

CHAPTER 11 – THE UNIVERSAL SPEED LIMIT

Ed – Now. There's a question you asked early on that we couldn't answer at the time, but now we've got all the information we need to answer it.

Stu – Really? I've forgotten. What was it?

Ed – We were talking about nothing being able to travel faster than the speed of light. We imagined your train to be travelling at 90% of the speed of light, and you running from one end of your carriage to the other, also at 90% of the speed of light. We said it would be impossible for me standing on the platform to see you going passed at 270,000 + 270,000 = 540,000 kilometres per second, because that would be faster than the speed of light.

Stu – Oh yes, I remember! The question was, if it's not 540,000 kilometres per second, then what speed is it?

Ed – That's right. So let's work out the answer now. We have three different speeds to consider, i.e., the speed of the train, the speed at which you experience yourself to be running in your carriage, and the speed at which I see you running from my position on the platform.

Stu – OK. So let's make "V" the speed of the train as usual. And since we've always added a little "*m*" for things in my private universe, let's make "V^m" the speed at which I experience myself to be running. What shall we call the speed at which you see me running?

Ed – Since it's the combined speed of you and the train, let's use the "sigma" symbol, i.e., "ε", since sigma is traditionally used when things are summed together.

Stu – OK, so then it will be "V^ε".

Ed – Right. Now, since "speed equals distance over time", for me as bystander the speed at which I see you running will be distance in my private universe divided by time in my private universe, so:

$$V^\varepsilon = \frac{X}{T}$$

Figure 81

… and knowing from our derivations of the Lorentz transformations that:

$$X = \gamma(X^m + VT^m) \quad \text{\large ⑥}$$

Figure 18

$$T = \gamma\left(T^m + \frac{VX^m}{C^2}\right) \quad \text{\large ⑫}$$

Figure 32

… then:

$$V^\varepsilon = \frac{\gamma(X^m + VT^m)}{\gamma\left(T^m + \frac{VX^m}{C^2}\right)}$$

$$V^\varepsilon = \frac{X^m + VT^m}{T^m + \frac{VX^m}{C^2}}$$

Figure 82

… and if we divide everything by "T^m", then:

$$V^\varepsilon = \frac{\frac{X^m}{T^m} + V}{1 + \left(\frac{VX^m}{C^2 T^m}\right)}$$

$$V^\varepsilon = \frac{\frac{X^m}{T^m} + V}{1 + \left(\frac{V}{C^2} \times \frac{X^m}{T^m}\right)}$$

Figure 83

Stu – OK, I get that.

Ed – Good. Now, what speed will you think you're running at in your private universe, which is moving relative to mine?

Stu – Well, that would be:

$$V^m = \frac{X^m}{T^m}$$

Figure 84

Ed – That's right! So if we substitute "X^m over T^m" with "V^m", then, for step sixteen, we get:

$$V^\varepsilon = \frac{V^m + V}{1 + \frac{VV^m}{C^2}} \quad \text{(16)}$$

Figure 85

Stu – OK. So now we can work out how fast I'd be

running according to you. It would be:

$$V^{\varepsilon} = \frac{V^m + V}{1 + \dfrac{VV^m}{C^2}} \qquad \text{\textcircled{\scriptsize 16}}$$

$$V^{\varepsilon} = \frac{0.9C + 0.9C}{1 + \dfrac{0.9C \times 0.9C}{C^2}}$$

$$V^{\varepsilon} = \frac{1.8C}{1 + \dfrac{0.81C^2}{C^2}}$$

$$V^{\varepsilon} = \frac{1.8C}{1 + 0.81}$$

$$V^{\varepsilon} = \frac{1.8C}{1.81}$$

$$V^{\varepsilon} = 0.9945C$$

$$V^{\varepsilon} = 298,350 \text{ km/h}$$

Figure 86

Ed – So you see – you didn't exceed the speed of light!

Stu – No, but I got close! What if we took everything to the maximum, with the train travelling at the speed of light, so that "V" equals "C", and me running at the speed of light, so that "V^m" also equals "C"? Then how fast would I be running according to you?

Ed – Well, there's no need to imagine you to be running at the speed of light, because we've already got something on the train that's moving at the speed of light!

Stu – You mean the light from my lantern?

Ed – Indeed! So let's work out what I'd experience for the speed of the light from your lantern from my position on the platform when the train is travelling at the speed of light:

$$V^\varepsilon = \frac{V^m + V}{1 + \frac{VV^m}{C^2}} \quad (16)$$

$$V^\varepsilon = \frac{C + C}{1 + \frac{C \times C}{C^2}}$$

$$V^\varepsilon = \frac{2C}{1 + \frac{C^2}{C^2}}$$

$$V^\varepsilon = \frac{2C}{1 + 1}$$

$$V^\varepsilon = \frac{2C}{2}$$

$$V^\varepsilon = C$$

Figure 87

Stu – Yes! That's exactly what we said at the beginning – we experience time and distance differently, but it happens in such a way that, no matter at what speed I'm travelling, dividing distance by time will always give the same value for the speed of light for both of us! And no matter how fast I run, I can never go faster than the speed of light!

Ed – Indeed! The speed of light is a fundamental property of the universe – a link between energy and mass as we said at the beginning, and the absolute speed limit!

Printed in Great Britain
by Amazon